SHACKLED

Joseph Lingard's true story of his time in a
prison hulk, transportation and travels in Australia.

An account of scant 19th century justice, brutality, deprivation and eventual freedom.

Denys Lingard

Text Copyright @ 2015 Denys J.C. Lingard
All Rights Reserved

By the same author:

The Beautiful Obsession (2012) - Fiction

Introduction

A few years ago I was in Derbyshire's Family History Centre at the County Library in Matlock, when I was told of a short book published in 1846 by one Joseph Lingard from a village near Chapel-en-le Frith in the county, describing his experiences. This was the locality where I knew many of my antecedents had lived from at least the 16th century onwards, so I was naturally interested, but nothing had prepared me for the extraordinary story that he related.

A questionable indictment for larceny, was followed by a farcical trial at Derby Sessions in 1835 where he was sentenced to be transported to serve a seven year sentence. Then nearly two years living and working from one of the infamous prison hulks moored on the Thames, before embarking on a five month long sea journey to Australia in conditions not unlike those experienced by slaves a century earlier, with over a hundred prisoners dying during the journey. His life in New South Wales whilst serving his sentence, and afterwards travelling around the colony still in its early days of settlement, before working his passage back home, make a fascinating account.

Joseph was clearly no writer and dictated his memoir to someone - probably a local journalist - who noted down the narrative more or less as it came into Joseph's mind and then wrote it up later to be published by a local printer.

Whilst I have faithfully followed Joseph's recollection of events and his impressions during these years, to make it easier to read, and to avoid too much repetition and too many archaic words, I have made some appropriate changes.   I have also researched contemporary and later accounts and records to set his story into a wider context of life and times between 1835 and 1846

Denys Lingard July 2015

Chapter 1.

In an introduction to his memoir dated April 1846, Joseph Lingard begins by addressing his readers:

"I think it necessary before any person attempts to publish a Narrative of his Adventures, to inform his Readers who and what he is; for much depends on his credit, whether he is worthy of belief or not; as some circumstances in such a Life as mine may astonish almost any person; and make him doubt the truth; but I do assure my Readers, that I do not intend to say one word in this Book, that is not strictly true, to the best of my remembrance.

I was born at Chapel Milton, a small village in the Parish of Glossop, in the County of Derby; about the year 1789. My father, Joseph Lingard, was the Clerk and Sexton of Chinley Chapel, and very highly respected in his situation. He maintained his family very creditably, until about the year 1800, when a very violent Fever visited our little Village and proved fatal to no less than fourteen or fifteen heads of Families; amongst whom was my lamented Father, (my Mother having died about two years and a half before), as well as two Brothers and two Sisters. By this calamitous event five of us were now cast on the World, to get our livelihood as best we could. We were instructed in the weaving of Cotton Cloth, which went well in those days, and we made our living by it. I did not think it would have been my luckless lot to be sent to New South Wales, nor do I think that those who read this little book would have thought so, but it has so happened, I here present to the world a small account of what I have gone through."

It is small wonder that Joseph wished to assert his honesty and truthfulness at the start of his story as he and his friend John Simpson seem to have been the victims of a plot led by a more powerful local figure to keep them from giving evidence as witnesses to his fraudulent activities.

The owner or tenant of a cotton mill - Bridgeholme Mill near the village of Chapel Milton - was a man called Elisha Dickens.   Late in the year 1834 John and Joseph who both had their homes in the hamlet of Bridgeholme Green near the mill were witnesses to Dickens and his mechanics dismantling machinery over a period of several days and, under cover of darkness at night, taking it and other items away from the Mill.   It seems that all this stuff taken from the mill was the property of Dickens and he vociferously claimed that the Mill had been burgled and that the robbers had taken away machinery, stock and equipment.   He complained to the local magistrate in the nearby village of Chinley, Richard Simpson, and gained his agreement that two constables could accompany him to carry out a search of houses in the neighbourhood of the Mill including those of Joseph and John.

So on the 31st of December 1834, Dickens with Constables William Ashton and William Porritt entered Joseph's house which they systematically searched and took away many articles which Joseph claimed had been his property for at least five years.   A drop-latch value sixpence-ha'penny (just over a modern penny) that Dickens had caused to be removed from a door claiming it was his property, seems to have aroused more ire than anything else in Joseph's recollection as he could date exactly the time and circumstances when he had come by it.

However, losing no time - justice was swift if not so sure at this time - Joseph Lingard and John Simpson appeared before Simpson the magistrate at Hayfield the next day January 1st, 1835 and both were committed to trial for larceny at Derby Quarter Sessions.

The following day Joseph was taken by coach to Derby County jail and five days later on Tuesday January 7th appeared before William Palmer Morewood of Alfreton Hall the High Sheriff of Derby and was tried jointly with John Simpson for "feloniously stealing, taking and carrying away on the 31st December now last past" the following:

One rack of the value of one shilling
One latch of the value of one shilling
One wheel of the value of one shilling
Ten steps of the value of one shilling
Ten bushes of the value of one shilling
One oven
One grate and other articles the property of Elisha Dickens -
a list which differs considerably from the written record of the charge, which, to say the least, seems to show some confusion on the day.

As there were forty prisoners, thirty eight men and two women, involved in twelve separate cases on trial in the same courtroom that day, very little time was spent on any one case. No record exists of the evidence given or questions asked and answered if there were any, not even a plea of guilty or not guilty although one assumes the latter. It was clearly summary justice on a conveyor belt system.

Some of the offences for which other prisoners were being tried that day included:

"Being in enclosed land in the township of Hassop, armed with a gun, for the purpose of taking and destroying game."

"Stealing at Bakewell, ten pieces of cord wood the property of the Duke of Rutland."

"Uttering base and counterfeit coin at Heanor"

"Feloniously stealing and carrying away from the premises of Sir Roger Greisley, Bart. at Drakelow, a quantity of boards and tressels."

It seems that all the prisoners that day were sentenced to be transported overseas to serve their sentence, most, including Lingard and Simpson, for seven years.

Joseph tells us "I was informed that my Prosecutor (Dickens) after sentence had been passed, sent home to inform his friends what he had done, and ordered them to get a goose ready for dinner on his return. Poor entertainment in my opinion, when his best sauce would be, the great exploit he had achieved in transporting an innocent man."

The next eight days he spent in Derby gaol until he relates "On the 14th I got breakfast at Derby, dinner at Northampton, and supper in Newgate Gaol, London. I was almost frozen to death on the top of the coach, it being a very cold day, and I being heavily ironed."

The following day he was taken to Woolwich and gained his first sight of the dreaded prison hulks on the Thames.

Little more than a year later in March 1836 his prosecutor Elisha Dickens was declared bankrupt, and it is rather ironic that the Bridgeholme Mill, that figured so prominently in the early part of Joseph's story should, later in the 19th century pass into the ownership of another branch of the Lingard family together with the Mill House and much of the rest of the village of Chapel Milton, staying in the family hands thereafter for over 100 years.

Chapter 2

When they hear the term "prison hulk" most people will recall the opening of Dickens' Great Expectations and Pip's dramatic encounter in the churchyard with the convict Magwitch who had escaped from a prison hulk moored off the Kentish marshes.

These hulks were for the most part de-commissioned "men o'war" or "merchantmen" from which all the masts, rigging and much of the superstructure had been taken away and the interior adapted to hold the maximum number of convicts.   They played a significant role in the English Penal System for over 80 years from 1776 to 1857.

 Problems with a burgeoning prison population in England were becoming apparent even before the end of the 18th century and more intensely on into 19th.   This was mainly due to the largely peasant population, many of whom were no longer needed in traditional agricultural work as a consequence of the Acts of Enclosure, increasing rent demands and land seizure by larger land owners, and their movement towards the larger towns and cities, in a vain attempt to find employment of some sort.

This swelled the numbers of destitute people in London in particular, leading to desperation, debauchery and crime.   At a time when the death penalty was prescribed for some 300 offences ranging from murder down to the theft of anything valued at over one shilling (5 pence today), and lesser crimes, resulting in long prison sentences, the main London gaols like Newgate, Marshalsea and The Fleet were vastly overcrowded.

It was the urgent need to get rid of this excess prison population that led to the introduction of overseas transportation for convicted criminals in a parliamentary act of 1717. The new colonies of Maryland and Virginia in America thus received hundreds of prisoners every year. There were no prisons or indeed any formal arrangements established to receive this influx of prisoners who were auctioned on the quayside of their arrival port such as Norfolk, Alexandria, or Annapolis to anyone prepared to accept the prisoner's indenture for the length of the prison sentence. As in most cases there was little or no pay involved for the prisoner's service, only provision of bed and board - such as it was. It was tantamount to slavery.

Few people in the American colonies were happy about what eventually totalled tens of thousands of petty thieves, prostitutes and the like being dumped on them. Benjamin Franklin for one suggested that the Colonies should "repay Mother England's kindness by exporting rattlesnakes to the British Isles, releasing them perhaps in St. James' Park." However, transportation to America was eventually only ended by the American War of Independence.

A few prison ships then embarked for Nova Scotia the West Indies and even Florida but the demand for indentured labour in Canada was low and slave labour was easily available elsewhere so the loss of the regular dumping grounds for hundreds of England's convicted undesirables could not quickly be filled.

Immediate action was called for and the Admiralty was prevailed upon to take a hundred convicts onto two old warships docked at Woolwich on the Thames.   This was initially a temporary measure, and met with a good deal of opposition but, despite that, the arrangement was regularised by Parliamentary Bills that confirmed the practice of using hulks as floating prisons, and utilising the convict labour on public works contracts.   Although other alternatives were explored over the next few years from building large new penitentiaries to shipping convicts out to fend for themselves in deserts or islands, it was decided to contract-out the establishment of a fleet of these hulks.

The first to be commissioned for that purpose was a ship called *Justitia* originally used to transport prisoners until 1772, when considered unseaworthy.   Then four years later she was stripped and refitted to become the first floating prison holding hundreds of convicts below decks, moored off Woolwich.   Others followed - at Woolwich and at Deptford on the Thames and at Chatham on the Medway, at Plymouth and Langstone harbour and some at other provincial ports.

Since the earliest days of the hulks being moored off the Woolwich Warren, prisoners were employed daily working as labourers on land clearing and building on the area which included the Woolwich Arsenal.   Others were employed working from small boats downstream from the hulks dredging ballast for the building work and also mud and sand to keep the channels clear.

Conditions were appalling and deaths of prisoners in the early days were running at between a third and a quarter of the convicts and these conditions remained relatively unchanged for the next forty to fifty years despite the efforts of reformers like John Howard.

In 1787 the First Fleet carrying 600 male prisoners from the hulks in England set sail from Portsmouth for New South Wales in Australia giving some hope to the despairing prisoners left behind. Many prison ships followed taking men and woman that had been confined to the hulks for many months or even years so that over 160,000 had made the journey to Australia or Van Diemans Land (Tasmania) before convict transportation ceased in 1867.

In 1835 however, the penal regime was in full swing and it was to the third successive hulk with the name *Justitia* still moored off Woolwich that Joseph Lingard was taken.

In his own words "On arriving, I found six hundred prisoners in the same condition as myself. My own clothes were immediately taken from me and others given to me - not very warm ones - so one might very well see through the cloth - and as still further accommodation, each prisoner had a pair of irons rivetted onto the left leg with links coming up to the knee. These were companions in the daytime and at night so we called them "Old Wives" as we were attached to them all the time.

There were seventeen classes in the ship - eight bottom, six middle and the remainder in the top deck, I was in the sixth class bottom deck which was supposed to hold forty two men when filled, it was eight yards long and five wide.   (Note: This would give each prisoner a space to lie down of about 17inches in width and six foot long in two lines with six men in the space between them.   Packed like sardines in a tin would aptly describe the class area.)

On each deck there was passage made of iron rails down the centre, along which the guards walk at night to keep all quiet. The classes were on each side of this. The captain, mate and their families lived in the cabin of the ship and the cooking was done in a part of the top deck."

The portholes were inevitably kept open throughout the year for obvious reasons, and Joseph comments that there was no fire to warm inmates against the keen cutting wind from the north that was often felt.

Charles Campbell describes it in his book *The Intolerable Hulks* - "Life aboard the hulks was a protracted horror for some and nothing better than an ordeal for others. But for the more sophisticated and predatory, it offered opportunities for the kind of exploitation of the weak by the strong that has characterised prisons for centuries. For all but a few, whose special circumstances enabled them to gain certain comfortable privileges, the prospect of a term aboard the hulks gave good cause for despair. Londoners would have tolerated hulk confinement somewhat better than those from outlying towns and countryside. Most of the London convicts had been slum dwellers, accustomed to privation, stench and close proximity."

Another vivid contemporary account from a prisoner, of hulk conditions reads:

"On descending the hatchway, no conception can be formed of the scene which presented itself. Nothing short of a descent to the infernal regions can be at all worthy of a comparison with it.

This was a new scene of misery to contemplate, and, of all the shocking scenes I have ever beheld, this was the most distressing. There were confined in this floating dungeon nearly six hundred men, most of them double-ironed; and the reader may conceive the terrible effects rising from the continual rattling of chains, the filth and vermin naturally produced by such a crowd of miserable inhabitants, the oaths and execrations constantly heard amongst them; and above all, from the shocking necessity of associating and communicating more or less with so depraved a set of beings."

Joseph describes his rations whilst on board - in the morning four ounces of ships biscuit and a pint of "smiggins" which he said should be made of oatmeal boiled in water and drawn out through a brass tap, however he said it was actually made of sweepings of some sort of corn that could not be identified.   He goes on "we had to "seeth" it through our teeth and then wipe away the dirt and rubbish from our mouths to make way for another spoonful, and sometimes it stank so, that we were obliged to hold our noses whilst we swallowed.   We had this allowance from the cookhouse in kits - six men to each kit.   We measured it out into our individual cans with our spoons as carefully as they weigh out their gold at the Mint in London.   The allowance given to a farmer's pig would have been a dainty one to us."

He went on "We came in at twelve noon to dinner.   Four days a week we each had about a pint of soup, one pound of bread (as heavy as lead and the colour of a brick) three quarters of a pound of beef and bone - many times the meat was no bigger than my thumb, and on the other three days the bread with a little cheese and half a pint of small beer."   At supper 'smiggins' again with any bread that happened to be saved from dinner time."

These daily rations seemed to have changed little from when they were formally specified in 1776 until Joseph's experience in 1835, although the quality was invariably poor. There were no vegetables or fruit in the diet leading to widespread scurvy as well as tuberculosis and typhus. It was not until the 1850's that eventually steps were taken to make some improvement.

Joseph resumes "We were mustered on the deck in the morning and gangs sent out into the Arsenal to different types of work, with soldiers, their guns loaded with ball and bayonet at the end, accompanying and guarding us until a return to the ship at noon for dinner.   My occupation for the first twelve months was to be in the wash house, assisting to wash the prisoners' shirts etc; an occupation which nearly cost me my life owing to my being continually wet.   Leaving work in that state to return to my lodgings in the ship, where there was no means to dry my clothes. I pulled them off wet and put them on again in the morning in the same condition.   At the end of twelve months I was afflicted with the black scurvy, from toe-end up to the hip quite swollen and black and unable to walk at all.

I was taken to the hospital, which was an old man o' war ship, used for that purpose and moored down river.   I was put in bed for three weeks and attended by two doctors.   I was not alone in this affliction, I had two hundred companions labouring under various ills, and dying on every hand.   I was in this truly dismal place for five weeks, then brought back to the ship again quite recovered. My occupation then was assisting to work the mud barges down the river and the canal according to the tide-times.   Two months after I had an attack of the ague but did not leave my work."

Joseph was particularly struck by the distress of the young women who were to be transported, he says "Often have I seen as many as fifty at a time brought down to Woolwich to be sent off.   One time in particular I remember there were twenty-five servants from 16 to 25 years of age who had robbed their masters.   They had been led astray by a set of most profligate young men acting in the capacity of sweethearts.   These poor deluded creatures had gone from one extravagance to another, till in order to carry on they were obliged to steal; for which they were leaving their native country and their friends for ever. Some were as beautiful young women as I ever saw in my life, and it was pitiful to hear their screams and moans. Such scenes would have broken a heart of stone.   In the time I was at the hulks eight vessels had gone to New South Wales with female prisoners only."

"Misfortunes seemed to haunt me" he continues going on to recount an incident on board the *Justitia* which resulted from his inability to read or write. A colleague of his in the same class named Oxley forged the writing and signature of another prisoner called Hill in a letter to the Hill family asking for money, and asked Joseph to smuggle it ashore and with the cooperation of a friendly soldier, Thomas Benny, who had it delivered.   Benny received ten shillings for his trouble and Joseph delivered the balance of thirty shillings to Oxley.   A day or two later Benny enquired whether Hill was happy with his money because he of course had read the letter. Joseph told him it was for Oxley not Hill and then realised how he had been duped and deceived. Someone persuaded Benny to report what had happened to one of the guards who in turn reported it to the First Mate.   Consequently Joseph was stripped naked and searched but no money was found on him.

The captain was away and until his return a fortnight later, Joseph was put into The Black Hole. He describes it as follows. "In this hole I had about a yard and a half to move in, it was winter time, and the wind which came through the stern ports was ready to cut me through.   I had no light but through a small iron grid over my head. My victuals were reached down at it three times a day and but poor fare indeed.   I remained in this miserable place, day and night till the captain returned.

I was then brought out on the quarter deck for examination before the captain and the first mate.   Oxley and Hill were also brought up.   Oxley denied that he had ever given me a letter and said that he knew nothing about the matter.   Hill said he should not have cared if they had sent for all the money his father had.   The first mate said that I had forged the letter myself, and the captain said the same and called me a scoundrel.   The Captain wrote to Hill's family for the letter to be sent to him, but being afraid of getting into trouble they said it was lost.   I was taken back into the hole again to find that I had a companion there, a man named Mason who had scaled the Arsenal walls and made his escape only to be retaken shortly afterwards.

We were both brought on deck again and sentenced to receive five dozen lashes each - Mason for running away and me for forging, as they said.   As the custom was to flog while all hands were on deck we were taken into the yard.   There Mason was stripped and tied up to the triangle with the Captain, First Mate, the Doctor, two guards and the Flogger close by.   The Flogger having stripped commenced, and gave Mason five dozen.

Standing by, I saw the flesh cut through till I could see the shoulder blades as white as a tobacco pipe!   Mason never made the least complaint, he was then untied, taken down and his clothes put on. I was then ordered to strip, though so bad of the ague I could hardly stand.

I was taken to the triangle and tied up.   The Captain told the Flogger "Do your duty."   This was a command that was not much needed as the Flogger had already told the prisoners that he would cut me through.

Then the Doctor exclaimed "Stay your hand Flogger" and walking up to me said "my good old man let me look in your mouth" which I did, then turning to the Captain said "If this man is flogged he will die at the triangles."

The Captain "He shall be flogged!"
The Doctor "He shall not be flogged."
Captain "I say he shall be flogged."

The Doctor "He shall not" then turning directly to Captain Hatton "Are you the master in this matter or am I?   I tell you that I am." Then addressing the two guards ordered them to take me down immediately, put on my clothes and take me to the ironing house where it was warm, and look after me.   Then, turning again to the Captain said "this old man has had the ague for four months and has not been fit for any work at all.   I have in the course of that time asked Lingard more than once to go into the hospital, but he refused, saying he was afraid to lose his work."

The next day Joseph was called on deck again where the Captain ordered him to be taken to the blacksmith's shop.   There a pair of irons weighing eighteen pounds in all were riveted onto each leg, with a chain from each passed through a ring in the middle to which a rope was fastened and tied round his waist.

He goes on "I was sent out to work in the Arsenal with as much rattle about me as a timber carriage has.   I remained ironed in this condition for three weeks, night and day and was obliged to get a man to lift my irons into my hammock, as I was not able to do it myself.   These were very cold bedfellows."

Captain William Hatton who had been in command of the *Justitia* for over ten years in 1835 and was to continue in that position for over ten years more, was neither efficient nor just in his work but managed to retain his position for that long period by a combination of bribery, obfuscation of officialdom, deft footwork and plain skulduggery.   Not until an enquiry set up by Parliament into conditions on the prison hulks sat in 1847 headed by the redoubtable Captain William John Williams, did Hatton get his comeuppance.

Charles Campbell comments "Although in answer to searching questions he demonstrated his usual skill in dodging richly deserved blame for a number of dubious practices, he could not escape having allowed his ship and convicts aboard her to become filthy beyond reason over the two decades of his command.   Moreover it was established that he had more than once bullied Mr Bossey (the Doctor) into the birch caning of mentally disturbed convicts and had had prisoners scourged with the "cat" without approval. He disputed this but was unable to produce records to support his testimony.   He was superannuated.

During the same enquiry Dr. Peter Bossey was also questioned by Captain Williams who in his report to the Home Secretary criticised Bossey's treatment of "lunatic" convicts, and blamed him for the very large proportion of convicts that were stricken with scurvy through not ensuring that vegetables were included in the convicts' diet.   When pressed on this point Bossey said "the subject of the dietary was hardly within my peculiar province"   Another area apparently not in his province were the conditions on the hospital ship "Unite" also moored off Woolwich.
Captain Williams commented that the ship was filthy from stem to stern and that the convicts who were her patients were "begrimed with dirt" and "infested with vermin."

After the enquiry Dr. Bossey, who also had a lucrative private practice ashore at Woolwich was relieved of his position as Chief Medical Officer for the hulks moored there.

Just after Joseph had completed his three week punishment a convict transporter ship the *Prince George* came up the Thames from Bristol - it was the twenty third such vessel to come to Woolwich since Joseph had arrived there.

Chapter 3

Since Captain Cook and his expedition had first set foot in Botany Bay in 1770 no vessel had visited the Australian shore, it being at that time, off the main trade routes to North and South America, India, Russia, China the Spice Islands and anywhere else that, at the time, traders and colonists wanted to go.   The reports of Cook and others were hardly encouraging as they commented that the land up and down the coast from Botany Bay was sandy and infertile and there appeared to be no pines or other suitable trees that could be used in shipbuilding.   Moreover they said that the natives they encountered had been aggressive, primitive and mostly unapproachable.

Under tremendous pressure from the consequences of jam-packed gaols and vastly overcrowded prison hulks and having considered and rejected other alternatives for one reason or another, the Younger Pitt and his government reluctantly agreed in 1786 that the transportation of convicts to Botany Bay was essential and parliament approved.

Almost immediately Captain Arthur Phillips a 48 year-old Royal Navy officer semi-retired on half pay and currently farming at Lyndhurst in the New Forest was commissioned to lead the expedition of this "First Fleet" to Australia and become by royal appointment "Governor of our territory called New South Wales." Not blessed by the charisma of a born leader Phillips was nevertheless a competent sailor, and a practical, level headed and pragmatic man who grew in stature as the demands were put upon him, and his experience grew.

In all the fleet was to be composed of eleven vessels including his flag ship *Sirius,* three supply ships and six transporters temporarily converted to carry prisoners. Phillips had to continually battle with the Admiralty on the inadequacies and complete absence in their plans of various essentials. Much of their planning seemed to be wholly based on equipping convict transporters travelling on a six week journey across the Atlantic rather than sailing for eight-months to other side of the globe. Lack of space, no portholes or sidelights, and in many cases little headroom for prisoners - one of the ships only gave four feet five inches - were only a part of the problem. Lack of anti-scorbutics to prevent scurvy, inadequate victuals to feed the expedition and even the lack of appropriate quantities of armament and ammunition were but a few of the constant string of complaints that Phillips, who had been given no hand in the original planning, felt compelled to make and have rectified by a slow, bureaucratic Admiralty.

Whilst he managed to get some things changed for the better, Phillips again had no say in which prisoners were selected for transportation in this First Fleet. Bearing in mind that one of the secondary aims behind getting rid of unwanted prisoners was the desire to get the convicts to set up a colony which it was hoped after a few years would become self-supporting, the need for those with experience in farming, market gardening, brick-making, bricklaying and other practical skills would seem to be paramount. However no thought was given to this and in consequence in a transport list that survives, with 190 male and 125 female prisoners, 56% of the male prisoners were either from the ranks of the unemployed or labourers, only 4 were brick-makers or bricklayers, 6 carpenters or shipwrights, 5 shoemakers, 5 weavers but there were no sawyers or fishermen.

However, by January 1787 the first convicts were loaded from the Woolwich hulks, men into *Scarborough*, women into *Lady Penrhyn*, but it was two months before the remaining prisoners embarked, and a further two months elapsed before the full fleet was ready. Then typhus broke out on one of the anchored ships on the Motherbank near Portsmouth so after these and other setbacks it was not until May 13 that the signal went out from *Sirius* to the fleet to get underway for Botany Bay. To take advantage of favourable winds the course was set via Tenerife, in the Canary Islands, Rio de Janeiro and Cape Town thus crossing the Atlantic twice on a journey of 9,400 miles before facing a final leg of 6,500 miles to Australia. It was a difficult voyage and conditions, particularly when they experienced intense heat for several weeks in the doldrums were almost intolerable. They stayed a week in Tenerife and a month in both Rio and Cape Town, where necessary work on the ships and replenishing stores were undertaken. The whole journey took eight months from May, 1787 to 20th January, 1788 when the fleet dropped anchor in Botany Bay.

Considering the rigours of the voyage in which the largest number of people ever (736 convicts, and almost as many officers with their wives and children, seamen and marines) were conveyed over a vast distance, the fleet had only suffered the death of forty-eight people.

It was almost exactly fifty years later on the 10th December, 1836 that Joseph Lingard went aboard the *Prince George* to follow the same journey and with the passage of the years very little had changed for the better. He described what happened on the day of embarkation.

"The next morning after breakfast the command was given to 'Stand round to muster.'. All prisoners then stand in the greatest suspense, each with his can, spoon and hammock in readiness. The guard with the Bay Book in his hand unlocks the gate and calls out those prisoner's names who have been chosen to go abroad - I was one.   Then to the blacksmith's to have the irons "struck off" and a pair of Bay Irons put on.   We were then stripped, washed and given a fresh suit of clothes, before being marched 100 strong down to the launch, rowed out to the ship and embarked for New South Wales.   It was the twenty-third transporter that had come to Woolwich since I had arrived.

Having taken our situations in the ship we sailed to Portsmouth to receive the remainder of our cargo namely 150 more prisoners, then after considerable delays, with the pilot on board we eventually sailed till we came to the Downs.   There a great storm arose and we were driven back to Sheerness for shelter losing an anchor and breaking a cable.   So severe was the continuing storm that the prisoners were locked down below, and the captain and doctor ordered a lifeboat to be lowered so that they could go ashore, until reprimanded by the pilot who apparently said to them "There are more lives than yours in this ship and I shall not leave her as long as there is a plank in her. It was now Christmas Eve."

When daylight appeared we were within a few yards of a sandbank and the water was coming into the ship by tons and our beds were swimming up and down between decks.   We hoisted a flag of distress and were visited by the bum boat with provisions and a steamer carrying a new anchor and cable.   We lay at anchor for a week."

He went on to tell that during the storm "that 'half-rocked' fool of a doctor - a Northern Irishman - fired two distress signal guns.   The natives of Sheerness thought immediately that we prisoners were all breaking out and came with two gunboats loaded and ready to fire into us. Had not the captain, very opportunely, seen them and demanded to know what they wanted we might have been sunk. It seems that the firing of two signal guns signified a prisoner breakout and they were responding.

"The next day we weighed anchor and sailed for Torbay where we laid until January 14th our captain and doctor having gone ashore. By eight o'clock in the evening we got a fair wind, they immediately came aboard and we set off once more for New South Wales.

We were all now taking a farewell view of Old England most of us for ever.   What lingering looks were cast towards the shore to catch another glimpse of our native land.   Many were the tears shed, and many were the thoughts of those we were leaving behind."

In crossing the Bay of Biscay Joseph says that there were "many a clumsy thump" as the voyage was very rough for some weeks with heavy squalls and gale force winds.   Most of the ship's crew were sea-sick and as the prisoners were in such poor condition before coming aboard many were dying daily.   Their bodies were merely thrown overboard.

He comments that the doctor proved to be a real tyrant and there were daily floggings on the main deck.

In perpetual discomfort, with a rigorous disciplined regime, the *Prince George* continued on its way until they reached the equator or "the line" as it was termed. It was February 20th and the sun was so hot that he claimed that one could not stand on deck unsheltered. "About twelve o'clock at night" he relates "the captain with his speaking trumpet called Old Neptune and his Lady to come aboard and was told they would wait upon the captain the next morning. They appeared on deck in the morning in their carriage which was drawn four times round the deck by two sailors. The prisoners were all brought on deck to see the sport, some in the long boat, with others on the forecastle and bulwarks. Then all was got in readiness for the shaving.

 Neptune and his Lady sat on a platform between the main and quarter decks and nearby a sail held by its four corners was filled with water.   All who were on board except prisoners who had not crossed the line before must be shaved or pay.   There were thirty seven people who in turn mounted a ladder to the stage and sat down. The barber's clerk tied a handkerchief round his eyes and then with a bucket of grease, tar and soot and good large brush gave each in turn a great lathering with the mixture.   Then came the barber with his razor made out of an old iron hoop a foot long, with teeth like a saw.   He began to scrape and every one drew blood in copious streams, cries and resistance were in vain as he scraped away until finished. Afterwards each of the poor victims was thrown backwards into the sail full of water where he was roughly received by the Bear (a strong sailor) who gave him a good ducking, and after half drowning him set him at liberty for which he was no doubt truly thankful."

He goes on "the day being very fine and hot, thousands of flying fish were moving about in all directions, beautifully coloured dolphins were also there, as was a sperm whale blowing on every side. A shark too was on the lookout for his prey and porpoise fish might be seen by the hundred. On upon the whole we had a very pleasant day of amusement.

The next day we resumed sailing along in our former doleful condition, on many occasions suffering privations for small offences, such as our provisions being withheld and an occasional flogging. Our regular diet was in the morning a pint of cocoa and a little biscuit, the same in the evening. Three days a week we had a little pudding and a piece of 'Salt Junck' about the size of your thumb, the other days a little pork and peas soup."

They came to the Cape of Good Hope but did not put into harbour. The weather was cold with some snow. On rounding the Cape, as apparently was the custom as a help against what Joseph describes as 'black scurvy' which affected most prisoners to the extent that most were unable to walk, each prisoner was given a small portion of port and lime juice. Joseph ascribes this condition to "abuse and tyranny" on board and whilst the treatment that was meted out did not help, the cause of the scurvy was undoubtedly a lack of fruit and fresh vegetables. He describes his own state "I was black from foot to hip and all my teeth were loose in my head." The weather soon became very rough, the sea running very high so that they were tossed about until the prisoners were almost senseless. After a period of calm he goes on.

"On the 4th May, we had a sudden squall commencing about 4 in the afternoon that took us unawares, with all our sails, stinsails, main jib and flying jib set. Before we could take in any sail a storm of wind and rain arose, ripped up all our sails like so many ribbons. It also took away our main jib and boom so that we were quite disabled. About 8 in the evening we were going to bed when there came another terrible blast of wind which carried away the starboard side of the ship's bulwarks. Our hatches were battened down whilst the storm roared like thunder, and such was the volume of water we were taking on board the sentries in the hatchways were nearly drowned as the water poured past them in torrents and on down between decks so that our beds and bedding were swimming. Those who were able were baling the water out using the cocoa kits whilst the pumps were going as fast as possible throughout the night. When we moved about we had to take great care for fear of being knocked to pieces.

At one point the ship's carpenter came below to tell us that to save our lives we should do all that was possible to help or we should all go to the bottom. We stuffed the ports with our beds and blankets, some men were crying, some praying others cursing and swearing. At midnight we all supposed that we were going down but the hand of the Almighty preserved us."

"The next morning, "he said "the main deck was strewn over with yards, sails, blocks, ropes and broken timber. The storm had abated very little and carried on much the same until 4 o'clock the next morning when it much improved. One of the sailors went aloft and gave the pleasing intelligence that he could see the lighthouse at Sydney Head."

Joseph says in common with many others he gave thanks for their deliverance in escaping through the great storm without loss of life. The sailor who had brought the glad tidings was presented with a bottle of brandy by the captain for his good news.

The *Prince George* entered Sydney Harbour that same day and the next morning the pilot came aboard and on a beautifully calm day they made their way in to drop anchor opposite the town. It was the 8th of May, five months since leaving Woolwich.

Chapter 4

The arrival of Captain Arthur Phillips and the First Fleet at Botany Bay in January 1788 and their experiences have been chronicled in detail elsewhere and my only reason for summarising this account is to point up both contrasts and similarities in Joseph Lingard's story some fifty years later.

It will be recalled that the only information that the new arrivals, Phillips, his officers and men, had of the land, the flora, fauna and the people, was that from Captain Cook's log and others who landed with him in 1770.   It proved to have little value.

It was quickly apparent that there could be no settlement at Botany Bay.   The expected rich grassland that some of Cook's officers had reported with deep healthy soil and areas of woodland turned out to be scrubland with little depth of stony earth for cultivation and the trees were all large eucalyptus with their grey-green foliage that provided little shade.   Moreover the bay was open and unprotected from the Pacific rollers and the heavy swell and shallow water near shore made it difficult for anchorage.   Perhaps more compelling still was the fact that there was very little fresh water available in the locality.

A small expedition set off to find a more suitable location for the settlement and one was found a few miles north in a superb natural harbour with much of the attributes that were sought.   It was immediately called Port Arthur but later re-titled Sydney in honour of Lord Sydney the Home & Colonial Secretary to whom Phillips reported "our satisfaction of finding the finest harbour in the world, in which a thousand sail of the line may ride with the most perfect security."

Work started on erecting tents and huts not least to accommodate the female prisoners who did not disembark for a fortnight.   On the day that this was finally accomplished and the female prisoners carried ashore, a tremendous storm blew up, tents blew away and the whole encampment became a rain-lashed bog as Bowes-Smith (a naval officer) describes it and went on "The women floundered too and fro, draggled as muddy chickens under a pump, pursued by male convicts intent on raping them." They were joined in the chase by rum-sodden sailors and Bowes comments "It is beyond my abilities to give a just description of the scene of debauchery and riot that ensued during the night."

The next day all were assembled ashore with a band of the Royal Marines in attendance to hear the Governor's Commission read by the Judge Advocate.   Governor Phillips was sworn in and in his address harangued the prisoners.   There would be no repetition of the previous night's events, any prisoners going to the women's area would be shot.   Cattle duffers and chicken thieves would be hung.   Those who did not work would not eat, and much else in the same vein.

Then work began, to clear land to build and where possible to till the soil and plant seed.   However the prisoners were ill-equipped to do any of the necessary work lacking both tools and ability, so that progress was tediously slow until skills were learned.   It should also be remembered that at this point few if any of the prisoners were interested in staying on for a future in Australia, they were not there as colonists and their fervent desire was to get away from the place as soon as they could, so an interest in building or farming was rare.   An attitude that was also widely shared amongst the naval personnel.

Prisoners were treated harshly and discipline ensured a minimum of trouble, flogging and occasional hangings saw to that.

Relations with the aboriginals became cautiously tolerant with curiosity figuring more often than aggression.   The naval officers and others felt that they exemplified a type of "hard primitivism" and there was no desire to exterminate or enslave them particularly as at first they posed no threat.   But nevertheless many were to be destroyed by cholera and influenza germs from the ships.

Phillips made the position on the treatment of the aborigines clear. Everyone arriving on ships had to "Conciliate their affections, and enjoin all our subjects to live in amity and kindness with them, and punish anyone who harmed them." However the prisoners' view was rather different as they looked with some envy upon the blacks with their complete freedom to wander about and their easy, seemingly indolent undisturbed way of life, at no one's beck and call.   So relations between them began badly and soon worsened. Many aborigines were skilled in making tools, clubs and spears which they valued highly because of their importance in sustaining their way of life in hunting and fishing, however they were often left lying about, providing easy pickings for prisoners who came across them to keep or sell to sailors as souvenirs.

As a consequence from to time to time convicts were found speared and mangled in the bush and on isolated work sites, the perpetrators all melting away making apprehension an impossible task and, as convicts never had the means or the opportunity to wreak revenge themselves the enmity simmered particularly as Phillips refused to authorise any retaliatory action.

Relations between convicts and the military also deteriorated as the latter resented the fact that punishments under military law were often stricter than those applying to prisoners and furthermore they did not like to feel that the food rationing that was endured as a necessity of preserving supplies were practically the same for military personnel and prisoners alike.   To the marines Phillip's even-handedness was bias.

The early months of the settlement across the spectrum of the personnel involved were both miserable and unhappy.   As Robert Ross the marine major whom Phillips had made lieutenant governor wrote: "in the whole world there is not a worse country.   All that is contiguous to us is so very barren and forbidding that it may with truth be said that *here nature is reversed,* and if not so, she is nearly worn out......If the minister has a true and just description given to him of it, he will surely not think of sending any more people here."

But, painfully slowly, work was done in building and cultivation, with frequent set- backs due to unhelpful weather conditions and inadequate equipment.

As Robert Hughes in *The Fatal Shore* writes "The hateful equaliser of all was hunger."   The First fleet carried enough supplies to keep its passengers alive for two years.   The weekly rations for all included beef, pork, dried peas, oatmeal, hardtack, butter and vinegar.   Theoretically an adequate allowance - in practice it meant scurvy, and most of the meat was bone and gristle.

By October 1788, Phillips, ignorant of when the first supply relief ship would come, had only one year's supplies in hand and decided to cut some of the rations and sent the *"Sirius"* commanded by John Hunter to Cape Town for provisions.

He returned in May 1789 to much relief, refitted and loaded with wheat, barley and flour, enough supplies for four months.
However in Cape Town Hunter heard nothing of any relief ships and there was no further news throughout 1789 and on into 1790.   The settlement was in despair and suffering a slow starvation.
Another disaster followed when *Sirius* despatched to Norfolk Island with her tender *Supply,* struck a reef off the island and sank.   All the crew and company including convicts were saved but all were now cut off from any means of communication and swelled the small settlement there that again was struggling to survive.

Phillips was forced to cut rations again at Sydney Cove and step up the punishment for food theft.   Morale was at its lowest as everyone suffering from hunger and malnutrition became apathetic and resigned to their fate.

Relief eventually arrived on June 30th 1790 with the *Lady Juliana* eleven months out of Plymouth, with supplies and news - of the French Revolution for example - and also the reason why no supplies had come earlier.   The *Guardian* en route to Sydney laden with two years supplies struck an iceberg, managed to limp into Cape Town where she was abandoned, losing all supplies. She would have been due in Sydney Cove many months earlier.

Less welcome however were the arrival on board the *Lady Juliana* of a further 222 women prisoners. They at least were in good health whereas on the other ships composing the Second fleet of the thousand that embarked, more than 250 died on the voyage and many of the remainder were very ill.

Phillips responded angrily to this blatant dumping of invalids from the hulks and prisons in a letter to London stating that if this practice continued "it is obvious that the settlement, instead of being a colony that will support itself, will remain a burden to the mother country."

It was clear to Phillips by 1791 that the colony could become self-supporting but not, he emphasised, on convict labour alone.   The prisoners had no incentive to work, they were, he said "flaccid" rather than rebellious, and the only way to go forward was to have "farmers and emigrants used to labour who could reap the benefit of their own industry."

They began to come in increasing numbers and by 1792 the colony was self-supporting.

Fifty years later when Joseph Lingard arrived in Sydney harbour on the 8th May 1837, building and land cultivation had transformed the town and new thriving settlements had been established. Elsewhere in New South Wales like Newcastle a couple of hundred miles up the coast, Melbourne and Adelaide to the south, with Launceston and Hobart on Van Diemen's Land, and in Western Australia, Perth, Albany and Freemantle.   However only a few miles inland the country was still for the most part unknown.

Joseph resumes his story on disembarking. "On the 9th. twenty seven of us went ashore.   The people had to meet us with hand-carts to convey us to the hospital, we were so weak and bad of scurvy through hard usage and we were washed and put to bed. John Simpson was one of the twenty-seven and the remainder of the ship's company went to the barracks.

However, they began to bring more and more prisoners from the barracks to the hospital every day and people were dying on every hand. As near as I can estimate we lost in our voyage from England and on arrival in Sydney around one hundred.

I remained in hospital five weeks and was then removed to barracks and put to work, though not even half well. We went to work in gangs with two or three overseers, forming new streets and other improvement in the town.

We were called in a morning at 5 o-clock for breakfast, taken in a large room off the yard called the mess-house. Kits were provided of Indian corn, ground and boiled in a copper. We were called in six at a time, one of the six has a kit handed to him around which the six stand, those who have no spoon ate with their fingers or went without. At six o'clock we went out to work and kept at it till half-past three in the afternoon. We had dinner at four; six men to a kit again - half a pint of soup and a little beef to each man, with three loaves to be divided - sour as a crab, solid as clay and the very colour of a new born brick. This was our regular food every day without any change. Two meals a day, no more! We went to bed at nine."

He went on "The Barracks is like a large factory, we lay two rows in each room, up to the wall on each floor. We had plenty of company in the night as rats came by the hundreds; they even came into the bed, they crept in at the breast and out at the feet like a pack of hounds, biting at our noses and ears the night through. If we had the least bit of bread under our pillows (and this was seldom the case) the rats would have had it.

There were seventeen hundred of us in the Barracks, plus two floggers and a counter who did no other work but regularly attend the flogging department and they worked full time.   There was a yard and two pairs of triangles for the purpose, and flogging went on every day except Sunday - I have known as many as twenty-nine flogged in a single morning till their backs were as red as a round of beef. I have seen men come in from their labour and have to bathe each other's shirt out of the wound where it had grown in during work.   Worse, I have witnessed boys nine or ten years of age flogged o'er the breech till I could have laid my hand in the wound - the children crying out for their mothers, who alas were seventeen thousand miles away.

This was how life was, whilst I was in the Barracks, which was for about a month.   Simpson my friend, came out of Hospital to join me two days before I left since when I have never seen him again."

Circumstances then changed for Joseph as he writes: I was signed out to a Mr Hannah, Quartermaster of the 4th Regiment - "The King's Own." Seven of us were sent up to a place called Parramatta (30 or 40 miles inland from Sydney towards the Blue Mountains) to a Captain Moffat's.   We stopped there for three weeks awaiting some cattle being brought up from Port Stephen.

On 17th August we started our walk to a place called Manara some three hundred miles distant.   We carried our beds and blankets on our backs as well as a week's provisions as we drove the cattle. Every night we lay out of doors and the weather was very cold. We made slow progress as the cattle were already tired when we started out.

In order to bake our bread on the way, we got a sheet of bark from a tree to knead our dough in and baked it come morning in the red hot ashes of the overnight fire. We carried a kettle, frying pan, drinking pots and provisions in bags, which we replenished at stations along the road every week. When we arrived at Manara, a journey of twenty seven days, Captain Camel our master was already there.

With the fatigue of the journey I was quite a done man. The next day we were examined, some were made shepherds, some watchmen, all were put to some task. I was set to thrashing corn, a job that lasted three months, then we came to the time for both harvest and sheep shearing. When all this work was done the man who looked after the kitchen was appointed to go to Sydney with the wool in three dray loads weighing about six tons. Due to my good behaviour I was put in the kitchen during his absence to wash, bake, cook and milk etc. for Captain Camel, and although after three months the man returned from Sydney I had managed so well in his absence that the master ordered me to continue my work in the kitchen.

Four months after this Captain Camel gave me the store keys, to give out rations to the men, I had nothing locked from me whatever. I continued in this situation until I was given complete charge of all the men on the farm, to put them to such work as I thought proper." He adds "In consequence of my good behaviour, I gained the esteem of both master and men."

Chapter 5

Joseph Lingard devotes a section of his little book to his observations on the way of life, customs and attitudes of the Australian aborigines, his keenness to learn more about them extended as far as learning some of the languages that the local tribes spoke in the area around Manara.

When the First Fleet arrived in Australia in 1788 there were probably around 300,000 native Australians living in that vast territory, most of these were concentrated in the coastal areas where there was more food and a higher rainfall but even here there was a density of population of probably no more than 3 people per square mile. For the original Australians the arrival of the convicts was an unmitigated disaster.

Almost from the start the convict population looked upon the aboriginals with enmity, they saw the preferential treatment that the law provided for the protection of the indigenous population, and that, together with their freedom to wander about at will, strongly contrasted with the lot of the convicts, and dislike led to hatred.

For their part the aboriginals seemed to have despised the convicts when they saw them labouring under conditions, driven, harried, kicked and flogged that their own pride would never have accepted. Clashes between the groups were inevitable even in the early days of settlement and worsened as time went on.   After prisoners had served their sentences and became freemen they still resented and feared the natives - the one group blamed "black resistance and treachery," the other "white retaliation and murder".

Early on, even armed with a gun, the settler was at a disadvantage to an aboriginal armed with a spear especially in the close bush where he could not load and fire quickly enough and few were good shots anyway. The aboriginals had no political or social cohesion and with over five hundred languages spoken across the Australian continent there was a continuous state of tribal warfare aggravated by their nomadic existence and poor communication.

One formidable difficulty that challenged the establishment of good relations was the degree of understanding between the two camps - native and white. Complex and ancient ideas about territory were embedded in the aboriginal collective memory and oral traditions, they had no concept of actually owning land but rather it being in the possession of mystic ancestors, the here and now just didn't matter. Such ideas were beyond the comprehension of the whites, and they paid them scant attention.

Legislation introduced in New South Wales in 1836 acknowledged that aboriginals were too few in number and too poorly organised to be considered free and independent tribes, also they had no specific rights to land as they were nomads. Nevertheless they were to be protected, and the killing of natives would be punished with "the utmost severity of the law." However the same legislation said that "any settler is not to suffer his property to be invaded or his existence endangered by them." As aborigines had no grasp of the basics of English Law and could neither be prosecuted nor serve as witnesses the scales were heavily tilted against them and the settler had a relatively free hand to dispossess or move on the aborigines as the need arose.

Between 1800 and 1830 better armed settlers moved further round the coast and inland over the Blue Mountains and north to the valley of the Hunter River.   As they pushed further, pitch battles were often fought between local tribes and the whites resulting invariably in heavy loss of life among the aboriginals who then also suffered from the introduction of stock to graze where formally there were kangaroos and other game to hunt.   It has been estimated that the death toll from this long frontier war was 2,000 to 2,500 settlers compared with 20,000 natives.

The result of this long bitter warfare depleted some tribal groups so much that the remnants became "station blacks"- who lived for the most part on handouts and occasional work.

Such is the background to Joseph's position at Manara.   He writes "The blacks, natives of Australia often visited me in the kitchen.   They go about in tribes and in summer time they all go naked - men, women and children.   They have no names amongst them the black man will call his wife Jen, and she calls him Blackfellow, and the children pickaninees.   When a child is born near any of the stations, whatever name any white man gives to the baby it will thereafter be known by that same name.

The blacks ramble about from one station to another and into the bush.   They have no system of labour, or any habitation.   At night they make a fire and sleep round it.   Their stature is low, generally the women are quite light and straight and some very comely with beautiful black hair.   They are thick lipped with a complexion a shade or two lighter than copper.

In winter, which is not so severe as in England, having never fewer than ten hours of sun, and in summer not more than fourteen, they would come in large numbers into the kitchen.   I have had as many as twenty women in at once, all naked as ever they were born, and men too, many times.   Every tribe has a chief or king, the wilder ones would wear a stringed dead man's hand or jawbone around his neck whereas those who live near stations and have become more cultivated have been presented with a medal that the chief hangs there instead.

When in the bush the women will make a fire at a convenient place whilst the men are employed, if near a river, spearing fish, ducks or wildfowl, or inland up in the trees catching opossum or squirrels, or on the ground collecting grubs, ant eggs, yams, roots or kangaroos. When they have gathered or killed enough raw food, it is brought to the fire to cook.   Then the women and children withdraw whilst the men continue to roast the produce skins, feathers and bones and all, and then eat till they are satisfied.   Then the women and children draw up and eat whatever is left.   If the weather is wet the women will gather sheets of bark and fix them on a few forked sticks to provide a crude shelter otherwise they have nothing to cover themselves with day or night.   When in travel the woman has to carry all the luggage, in addition to her infant as the man will carry nothing but his implements of war"

Joseph goes on to describe the aboriginal's customs.   "A father or a mother can give a daughter away to whom they think proper - whether she be willing or not. A brother can also give away his sister in a like manner.   If a man kills his wife they have no law to punish him.   I scarcely ever saw a married woman who had not got six or seven cuts in her head several inches in length and very deep given by her husband with a tomahawk.

I am sorry to say that aboriginal women will often murder their own children."

He gives several instances where this had happened. Particularly gruesome is his description of one woman whom he witnessed dashing her child's brains out against a gum tree and then throwing the lifeless body onto the fire. Such instances he comments were not at all uncommon. If a wife leaves her husband she will be beaten by the other women, if she does it three times then she will be taken into bush by the men and speared to death.

Joseph goes on "When a young man has arrived at maturity he is not allowed to have a wife until the front tooth of his upper jaw is knocked out. A day is fixed for performing this ceremony when a number of young men are taken into the bush and in turn a tomahawk is held to the tooth and hit with another. Then the man is ready for a wife when he can get one given to him."

The local aboriginals he says were usually able to count up to four in their tribal language and thereafter relied on fingers. For example if they wish to tell you of something that happened a few months before they will hold up their fingers to indicate the months that have passed but you will have to count them for yourself.

Every year at the same time, Joseph relates, his master, the men and most of the aboriginals from about the station drove the sheep to a place on the Bunbarlow river about five miles distant where they stayed for three or four days washing the sheep prior to shearing.

The second year he was left entirely on his own at the station and three aboriginals that he knew as Tommy, Charlie and Harry arrived in the kitchen in their war gear - hair tied up in a bunch on top of their heads, their bodies painted with pipeclay and raddle and spears and tomahawks in hand.   He asked them what they wanted and they said tea, sugar, tobacco and bullock.   Initially he thought that the master had sent them as he did occasionally if they were short on supplies, but their intent became clear when they jabbered away in their own tongue threatening him with death if he didn't immediately go and unlock the stores and give them what they demanded and some grog.

He realised that with nobody about he was in mortal danger so keeping his eyes upon them he moved backwards and picked up a heavy mundle - a two handed gripper for turning logs on the fire - swung round and hit the first man between head and shoulder knocking him to the floor.   He jumped over him and hit the next man on the back of his shoulder and brought him down, he made to strike the last man who was outside by this time but he jumped on a rock and was about to throw a spear through Joseph when he threw the mundle at him and ran the fifteen yards back to the house.   There he retrieved a loaded double-barrelled shotgun and ran back, to find Tommy was swimming away across the river. Charlie was standing on the rock holding his neck which was bleeding, with Harry still threatening nearby.   Joseph took aim at Charlie and hit him on the backside and he screamed and jumped away to hide, then he turned the gun on Harry and discharged the other barrel.   After this he needed to run back for his powder flask and shot belt but on return saw that all three had swum the river and were climbing the other bank.   He warned them not to try anything like it again otherwise he would finish them off.

Joseph describes some of the battles between tribes that had been fought nearby when he was on station. "They dress up in their war paint and face each other on the battle ground where a single spear is thrown into the camp of the other tribe to signal the beginning of the battle, then a single spear is returned before battle is joined in earnest. Spears are thrown thick and fast with shields guarding them off, followed by slinging bumrings (sharpened boomerangs perhaps?) which he says are capable of taking off an arm or a leg instantly. Then at close quarters they fight with striking weapons and shields he calls nulla-nulla and heelaman, and sometimes tomahawks. They go on fighting until one side gives in and the battle is over. Should any of the men have been killed, the winning side takes the women that belonged to them. If the women refuse they are killed."

He states "The victors cut away any flesh that they fancy from those killed in battle which they broil and eat, the fat they extract they rub all over their skin vowing that it will make them as two men." He tells that occasionally women have come into his kitchen bearing lumps of flesh and dead men's hands and promptly been driven away.

After the battle the winners have a great celebration called a Carrobbery with neighbouring tribes with whom they are at peace. They make rings of fire in an area of about two acres and in their full war paint dance and sing with the women as the audience, the celebrations often lasting for two or more days.

Chapter 6

Joseph worked for Captain Camel for three years, but the man who had signed him out originally it may be remembered had been a Mr Hannah who sold out his land about two hundred miles distant at a place called Malongla, to a Mr Hoskins.   Consequently all the men belonging to Mr Hannah were recalled from Mr Camel and went to another Hoskins station at Cambelong five miles from Malongla.

He says "I was the last to leave and with a sorrowful heart.   I did general work at Cambelong for about a week then was asked by the overseer, to take over the kitchen and a few days after that my new master handed me the keys to all the stores.   Initially I refused the responsibility but he pressed me saying - you must do it, for I have no one to trust when I am away from home, through the character that your last master gave you I can trust you with all I have got.   I agreed and all went well for three months."

Then after a big land purchase involving a number of stations in the area, Cambelong was acquired by new owners and Lingard and the other men found themselves back at Malongla and again he was appointed to manage the kitchen and stores and there he stayed until the time serving his sentence was completed on January 7th 1842.   He was a free man again.

His former master Captain Camel invited him to stay with him at Manara for as long as he wished, as his guest, and Joseph collected all his belongings from Cambelong and Malongla and brought them with him.

Joseph's main ambition was to return home to England - despite his treatment in the name of English justice - as soon as he could, as he retained a very sentimental attachment to "Old England" as he continually referred to it, and his book often quotes poems that reflect the way that he felt about his home land whilst away. However, although he was determined to make his way home in due course he was equally enthused to see as much of New South Wales as he could, so over the next two years he travelled extraordinarily widely in the largely inhospitable outback where settlements or stations were few and far between. It is worth remembering that in 1842 Melbourne had only been established in Port Philip Bay a few years before in 1835, and Adelaide in 1836.

In an attempt to summarise some of his expeditions I have been frequently frustrated by the writer's attempts to interpret Joseph's pronunciation of place names into spellings that bear scant relation to present-day names on a map, or indeed have changed entirely in the years between. Moreover, when Joseph was travelling New South Wales covered a much bigger area and some places he knew are now in the state of Victoria or the area around Canberra in Australian Capital Territory.

He was keen to learn all he could of the flora and fauna which was so different from that of England and on his journeys shot many species not just for food but to skin, stuff and preserve in many instances.   Similarly he collected aboriginal tools, weapons and other curiosities to take back to England.

He was staying at Manara, Captain Camel's head station between 300 and 350 miles west of Sydney near to the border of what is now Victoria when he took off on his first expedition.

He had planned to go with a man called Edwards, who had waited patiently for a year whilst Joseph finished his time so that they might travel around together and then return to England. However it was not to be. Whilst Joseph was getting together a cart and harness to take all his boxes, equipment and a plentiful supply of guns and ammunition Edwards rode off one morning to his shoemaker's about twenty miles away with quite a lot of money on his person. Edwards declared he would return the next morning, but as Joseph went on "I never saw or heard from him again. I reckon he must have been murdered by bush-rangers or by blacks for neither he nor his horse were heard of afterwards. This disaster fell heavily on me as we should have been pleasant company."

Joseph waited a fortnight for Edwards to return, meanwhile Captain Camel had gone away for two months on a visit to Golbourn only one hundred miles short of Sydney, so Joseph went on his own, to a place called Quadong seven miles away, where Camel had a sheep station. He writes "I was much delighted with the place it was a beautiful plain on limestone, about two and a half miles square, bounded on one side by a winding river which on leaving Quadong entered a narrow passage enclosed by almost perpendicular limestone rocks to a great height. Hawks of all description harboured here and the river was covered with all kinds of waterfowl, and I could take my gun in a morning and shoot just as many as I wished. Our hut stood about a mile from the river, where Mr Camel had a flock of about a thousand sheep, one shepherd and his dogs were with them in the day. Every night the sheep were hurdled up and guarded by a watchman with four dogs, each of which knew his own station on a corner of the hurdle and would be employed all night in driving off the wild native dogs - which are more ferocious than the wolf.

I have known as many as seventy fat sheep torn into pieces in the course of a single night by these dogs.

It was a delightful journey for me, some days shooting wild fowl, others taking hawks in the rocks or parrots in the bush. There were no snakes stirring at this time they had retired for the winter, but" he adds, "in the summer months there are many in great variety" which he proceeds to list ending with the death adder, a bite from which he said is always fatal.

Warming to his subject of wildlife Joseph then goes on to set out the myriad varieties of animals and birds - many of them unique to Australia that he had seen whilst in the country.

After seven weeks Captain Camel returned to Manara and Lingard was able to ask him whether one of his men, John Ashmore originally from Hartington in the High Peak area of Derbyshire, could be spared to accompany him on an expedition into the bush, and this was agreed. Having prepared provisions, equipment, guns and ammunition they set off with two natives and a big bull-mastiff westward across the Quidong plain.

He writes "After travelling some two miles in the bush we arrived at the "Deep Creek," a place where the width is only about four yards, I could easily throw a stone across, but the depth is so great that the descent down and ascent up the other side is a mile. Here the blacks gave us the slip and we lost them. We were set on going rock-wallaby hunting, they are a species of kangaroo but much smaller, hiding amongst the rocks during the day and out feeding at night.

We continued our journey through the bush until we came to a long slope which terminated at the Snowy River.   Here there were thousands of acres of nothing but barren rocks piled one upon the other with large caverns beneath hiding plenty of rock-wallabies had we known how to get them out.   The blacks having left us, our journey was lost, and we had travelled barely eighteen miles and as night was approaching I began to search for water as Ashmore gathered timber and made a fire.   Having made tea we ate bread and beef before making ourselves a shelter from tree branches it being very cold and dark and dangerous to venture out with the Snowy river below making a tremendous roar.   We laid down with our big coats on and both guns loaded, with the dog beside us, and being well tired fell asleep. Sometime in the night the dog awoke us barking, and then ran down the creek towards the river.   Something, we could not see what, kept making attempts to get to us.   I held the dog thinking to bring it nearer but I could never manage.   The dog was ferocious past anything I had seen before and Ashmore was almost frightened to death. At day-break we refreshed ourselves and after reviewing this dismal and doleful place returned home."

This early experience of his companion John Ashmore's character gave Joseph some slight doubts as to his suitability as a fellow traveller.   As he comments in retrospect "he was a great strapping fellow, but he had not the heart in him the size of a gooseberry when difficulties came in the way."

It was now June - the depths of the winter in Australia, and Captain Camel had two men going into the bush to split timber and Joseph offered to take their tools with his own provisions guns etc., in his horse drawn cart.

They were going into an area where he was assured there was plenty of game including flying squirrels, as the timber was so big and tall. On the second day they arrived in the bush a little before dark, tethering their horse to a tree with sixteen yards of new rope to give him a chance to find enough grass amongst the fast falling snow. They lit a fire using dead wood and then felled two trees which they stripped of bark, then after making a hole in the trunk of a standing tree used one felled tree as a ridgepole supported by the other against which they layered the bark on each side, and the remainder on the floor on which they put down their beds. The fire was in the front so they had a supper of tea, beef and bread in some comfort. He goes on "with a tarpaulin in front for a curtain we all lay in warm comfortable beds with a good large dog and two loaded guns beside us, but I could not sleep much on account of the screaming of the squirrels, opossum and other animals I could not name, as well as the howling of the native wild dogs which kept on throughout the night despite our own dog continually driving them away.

The next morning three of us set off in search of squirrels leaving Ashmore and the dog in charge of our horse and stores. We went about a mile examining the trees where we thought the squirrels and opossum were. The trees in NSW always fade at the heart first both in the ball and the branch, and these animals lodge in the hollow places of these faded trees. With the first tree we felled, a hollow branch burst open and there lay two fine squirrels asleep. We snatched them up by the tail and they were unhurt but we were obliged to kill them. We worked hard and felled several more trees before noon but found no more squirrels.

We went back to camp for dinner expecting it to be ready but alas, instead the High Peak man had lost the horse.

He had loosed him from one tree to tie him to another, the horse twitched the rope out of his hand and away he galloped and away after him went our cook, so we had to get dinner for ourselves. The three of us went out tree felling again in the afternoon and returned with two more squirrels, and found the provisions safe and untouched. It was two days before Ashmore returned - sooner than he expected as the horse had got the rope fast around a tree seventeen miles away so he was able to secure him and bring him back.

I was skinning and drying my animals when Ashmore and the horse returned, I tethered him to a tree whilst the man got something to eat. Then he led the horse away to find some good grass whilst he was tethered, but instead of doing so straight away he sat on a log smoking his pipe holding the end of the rope in his hand when the horse, knowing what sort of customer he had to deal with, twitched the rope from his hand again and away he goes once more with Ashmore following. It was fully fifteen miles before he was able to lodge at a station for the night.   He followed the horse's trail all the next day but could not find him. On the third day a stockman found the horse with the rope fastened completely around a tree in such a way that the animal was not able to get anything to eat. He took the horse to a cattle station run by Mr Thomas Warburton who recognised the horse as belonging to me and took charge of it. That same night Ashmore arrived at Warburton's, and the following day arrived back our camp once again with the horse which, needless to say, I took charge of."

Sometime before, Joseph had been invited to stay with a friend of Camel's called Liscome whose property lay some fifteen miles beyond Warburton's station, at the foot of the mountain, so he and Ashmore loaded up once more and set off, staying with Warburton for two days en-route. When they arrived at Liscome's station he was entertaining a Scottish whaler captain originally from Dundee. This Captain Stephenson had founded a station near Cape Howe between Two-Fold Bay and Ninety Mile Beach, where he and his family had lived for only three months. He claimed that it was the finest place that he had ever seen for sea birds and invited Joseph to visit him.

Chapter 7

Lingard stayed with Mr Liscome for about a month shooting and collecting his specimens, whilst Ashmore was working on the property. However the country thereabouts was so heavily wooded that Joseph had difficulty in moving his horse drawn cart very far which he found inhibiting.   So somewhat earlier than he had anticipated they loaded up the cart with provisions, powder and shot etc., and set off for Captain Stephenson's.

Joseph relates "We had nothing to go by but a marked-tree line for the whole ninety miles, that is, here and there a chip of bark taken out of the trees in the direction we were to travel in. There was nothing but mountains all the way, and so full of timber that we could scarcely get through.   The first of these mountains was called Morris's Mountain, it was twelve miles of a journey over it. When we got to about five miles from Bundi, we got lost and were obliged to turn round again.   This soft High Peak man sat down and actually cried, however we made our way back and in the evening arrived safely back at where we had started from that morning. Two days later we started our journey again.

John Ashmore said he would not come, however I prevailed upon him to try again.   With difficulty we got over Morris's Mountain that day to find trees of an incredible size.   We reached a small river which, if we followed it, we had been told, we should find a station further down on its banks and an hour before sundown it came into sight.   I was very tired having carried thirty pounds on my back and two heavy guns.   Weatherhead, the station overseer was away, but his wife who lived here with her children, a servant girl and one man, agreed that we could spend the night in the men's hut.

We joined the family for supper, which was a hearty meal during which the mistress asked if we were the two men going down to Captain Stephenson's, as he had stopped there the other night and told her he was expecting them soon, and that he was presently away at Browley. He had left word for them that he had brought two blacks with him who had marked the way from his home to Genore stockyard corner.

We slept very comfortably and thought of starting immediately but the mistress came to the hut and said we had better rest another day as the way we had to go was a very rough one, and in the morning she would send a man with us to put us on the right track as it was fifteen miles to the next station, seven of which were very rough. We agreed to accept her invitation and I took my gun and took a walk up the river and shot a few ducks for her.

I saw trees there of a prodigious height, I estimated about sixty feet and twenty five feet through the ball. Returning through the paddock I saw in one corner that something had been interred and asked the mistress what it was and she told me the tragic story. They had a dog who had the bad habit of running the cattle without orders, her husband was out with his gun one day and caught his dog in the act and chased him home. Once indoors the dog ran under the sofa to hide, her husband tried to drive him out using the stock of his gun, which went off accidentally and shot his eldest daughter who was looking on, dead upon the place.

> Now, there beneath the shady trees,
> In solitude she sleeps;
> And often near her mouldering dust,
> Her mother sits and weeps.

Next morning we commenced our journey, the mistress sent the man with us for about seven miles through great forests with ranges of mountains on each side.   We reached the marked tree line before he left us.   The trees were so thick that we had great difficulty in making our way through.   At last we reached the river where there were fowl of every description.   I was most surprised to hear in large numbers, the bell bird which sounded just like a lot of bells ringing, there was also the whip bird which makes a great noise exactly like the sound a coachman makes when cracking his whip, neither of these birds are much larger than a robin.

The marked line followed the edge of the river for a while then took us up the mountain slope.   About noon we unloaded our horse and tied him to a tree so that he could feed for about an hour, made a fire for tea and a meal, then rested a while.   Whilst we were there, a large group of kangaroos came by and our dog gave chase but they quickly left him behind.   We packed up and set off again.   About an hour before sundown we mounted a high hill from which we could see about two miles away two men ploughing with a couple of bullocks.   This was a station they called Wong-a-ra-bar.

We jogged down the hill and when about a mile from the station I fired a gun which made such a roar in the valley amongst the trees that I was quite taken aback.   Hearing the gunfire the men left their ploughing and hastened back home to get their guns as they thought they were under attack from the bushrangers.   We made towards the station where the whole Donald family from the Scottish Highlands - father, mother, five sons and two daughters stood beside the house.

I saluted them and asked if we could stay the night, they agreed but wanted an assurance that we had not come to do them harm. I explained that we were on our way to Captain Stephenson's and again they knew we were expected. They helped us unload, set a meal before us and we spent a very pleasant evening together."

In the morning they started off for Genore fifteen miles further on which was reached at sundown near a river which was tidal at that point. Again Joseph fired his gun to give warning of their approach and they were warmly greeted by the stockholder. Whilst they were talking two men from the sister station over the river alarmed by the noise of the gun into thinking that there was an attack by bushrangers came and joined them. After realising their mistake they stayed on for supper.

A couple of days later they made ready to start off when Ashmore turned difficult and refused to go, whereupon Joseph read him the riot act and forced him to proceed, so eventually they left on their journey which initially was, as he said "nothing but mountains and thick forest with no track except here and there a chip taken out of the trees." He was as before, carrying thirty pounds on his back and two guns, with his companion leading the horse, tomahawk in hand to cut back saplings to make a clear way forward.

At noon they reached the top of a high point still surrounded with thick forest and rocks, but they could see in the distance the wide ocean and ships.

"We descended into a valley by a stream where we unloaded, had a meal and rested for an hour, then amidst much grumbling and growling from my fellow traveller we loaded up and set off again.

We had not gone far before we had to unload the horse to get him across a creek, this was the first of five times we were forced to unload to get across hazards in the course of the afternoon.   We were overtaken by nightfall when it was too dark to follow the chip-line so I unloaded and hobbled the horse and made camp on a large flat area.   Ashmore was 'sadly in the dumps' and would do nothing.   After making a fire I went in search of water and went blundering on through the thicket for a hundred yards until I suddenly fell over a precipitous edge and down about three yards into a hole ending up to my waist in water, so of course I soon filled my pots and made my way out another way."

After a night when he says the screaming of the night birds and animals was truly remarkable, they loaded up and set off again only to find within a quarter of a mile they were forced to unload to get the horse through vines, shrubs and boggy land. It was, he comments a most difficult section to pass through and continues "Soon we came to a very soft bordered creek and we could not get the horse over.   Ashmore would do nothing to help and we had words and he threatened to cleave my brains with his tomahawk. I stood with my gun loaded and finger on the trigger and told him if he made any such an attempt I would shoot him dead, throw him into the creek and leave him there.   In a short time his anger abated and after much difficulty we managed to get the horse over and went on our journey through grass-tree flats, gullies, mountains and forest until I lost the mark line.

This had stopped some two miles from the station and we lost our way.   My companion began to cry at the thought of staying out in the open for another night, however we pushed on another mile until we heard the roaring of the sea.

On proceeding a little further we came upon a track where someone had been dragging saplings to make a fence.   Although we were very tired we pressed on knowing it could not be much further, but as the sun was going down I fired my gun which made a great noise.   Shortly afterwards we saw a man approaching which put Ashmore into better spirits, the man was soon followed by a woman and three children.   We saluted them and asked if we might unload and stay the night as we were very fatigued. The woman was much frightened thinking we were bushrangers coming to kill them but being reassured he let us go towards the huts.
The name of the place was Malacoota another property belonging to Captain Stephenson who, they confirmed was presently away.

On learning his hosts were completely out of stock of tea and sugar Joseph gave them some of his supplies, and after a meal and tea he and Ashmore settled down in the men's hut for a good night's sleep.

Early the following morning Joseph cleaned his gun and set off walking along the adjacent beach for several miles watching the sea as he described it "come foaming in against the shore like moving mountains."   He goes on "I soon came to a winding river that took its course round a narrow neck of land, on the opposite bank I saw two white hawks and in an endeavour to get closer I stripped off and accompanied by my faithful dog struggled to cross, succeeding at last but with great difficulty.   No sooner than I had landed than the beautiful hawks took off and crossed the river to the other side. I crossed back put on my clothing and pursued my journey.   Three miles further on the forest sets in quite near to the sea.
Kangaroos were sporting about in all directions, my dog had many a race but failed to capture one.   I fired my gun at one but it was at too great a distance.

A little further on I fired at a bird and killed it, my dog ran and mouthed it and thus spoiled it for which I beat him. He ran home and I saw him no more until that night.

In some places, close to the beach grows a kind of vine, which forms a beautiful covering overhead so closely matted together that it will keep the rain out for hours. About thirty yards from one of these places, I was loading my gun again whilst admiring the wonderful works of nature, when suddenly a black man popped his head from under one of these thickets, but on seeing me he instantly drew back. My dog having left me I was much alarmed as I thought there may be a whole tribe concealed here, so to run would be of no avail. Then the black fellow looked out again and I beckoned for him to approach. He advanced several steps then stood still staring at me, he was quite naked and held no weapons so I beckoned him again, he came forward to stand in front of me and stroked my gun talking to it in his own language. On casting my eye towards the thicket I saw another black man poke his head out and I beckoned to him too and he came to join his colleague. Cautiously I drew back towards the bush to get out of spear range in case there were more concealed in the thicket. At that moment two crows flew into a nearby tree and perched together on the same bough and both black men pointed to them with their fingers, I backed away a little and fired bringing them both down. The natives capered around them gleefully then picked them up, stripped off the feathers and hastened away to their camp. I loaded my piece again and walked in the same direction. Ten minutes later the two men appeared again with the two crows part roasted, they were obviously a great treat because they were tearing them to pieces with their teeth. I made bold to go into their retreat which was like a house inside.

Their implements of war were propped in one corner and the floor was strewn with all kinds of shells and fish bones, they had two wood fires burning.   After surveying their home and the surrounding country I proceeded home, the day being far spent."

Chapter 8

Captain Stephenson's station was near to a fine inlet of sea water with some small islands (Malacoota Cove?) where there was an abundance of sea birds of great variety, and Joseph kept shooting and skinning specimens for the next two weeks. During this time Captain Stephenson came home, and they spent a pleasant and convivial night together.

"One day" he relates "a lot of blacks came down the river Genore (Genoa?), men women and children, and I soon got very friendly with them. As there was an abundance of fish thereabouts I brought some hooks out of my baggage and taught the natives how to fish with hooks which pleased them greatly. They made lines for themselves out of tree bark, and went fishing in their canoes which were made of one sheet of bark tucked up at the ends. It was winter time and they built fires in their canoes on a bed of clods of earth, onto which they threw the newly caught fish to broil. At low water they would gather cockles, oysters and mutton fish which they would tread out of the mud, these activities would continue throughout the day, and every night they would bring us fish - as much as we could cope with.

One of the aboriginals made me a canoe as a present. It would hold three people and I went from island to island around the inlet, sometimes with a black man or a black woman to help row. Occasionally the Captain accompanied us on a shooting expedition. At night I would often return with a few brace of ducks for Mrs Stephenson.

As we were running low on supplies of wheat, flour, tea and sugar Captain Stephenson asked if I would agree that my man Ashmore could go with one of his men to Manara, a hundred miles away to obtain them, and I agreed.   So they set off to Manara with two pack bullocks and a bull. On reaching Liscome's at the foot of the mountain, Ashmore stopped and let the other man come back by himself.   I had given Ashmore the key of my box that was lying at Liscome's in order that he could bring me back a can of powder and other items, instead of which he lent all the money I had in the box to Liscome to carry him to Sidney.   I never saw the money again.

Every day we were waiting expectantly for the supplies to come, tea and sugar ran out and we were reduced to eating the siftings of the remaining wheat.

One day the Captain and I, taking a black woman as guide, went across the river and on to a wreck about ten miles distant, close to the Ninety Mile beach. The ship had been wrecked at night about two years before on some very high rocks en route for Sidney and was now scattered in a thousand pieces on the sand.   Eight women and seven men had been drowned and I could see some of their shoes and bones amongst the rocks.   Those who escaped were attacked and killed by the natives but not before considerable casualties had been inflicted on them by the ship's crew the evidence of the battles lay in the bones that were still there on the beach.   Those who survived made their way to Two Fold Bay. One story has it that some sawyers who visited the wreck to see what might be salvaged, found the body of an old gentleman floating in shallow water by the beach with a bag of five hundred sovereigns hung on his arm.

Although interesting in some ways, it was an uncommonly wet and uncomfortable day.

I had stayed at Captain Stephenson's about ten weeks, shooting, skinning and drying new specimens and latterly waiting impatiently for my man Ashmore to return and accompany me into the bush. I had my dog and horse by me but could not go by myself as spring was drawing on and snakes were making their appearance in large numbers, some of them very big.

As Captain Stephenson and his man had to go to Genore to fetch some seed potatoes I decided to travel with them and took leave of his wife and family. She told me that she was frightened when I came but was really sorry now to part with such a good friend, she wept and the children clung round me as we bade farewell - I suppose - for ever. We struggled over the thirty five miles with three horses and a pack-bullock but despite the difficulties reached Genore that evening I stopped there for two days then finding the stock-keeper was travelling to Nan-gutty for some calves the following day I decided to travel with him part of the way. I bade the Captain farewell and travelled with my companion the sixteen miles to Wong-a-ra-bar station where the Scottish family Donald lived. Whilst the stock-keeper went on to Nan-gutty I stopped with the Donalds for a day before joining some of them who were going to Manara and they proved very good company. The Donalds like most of the families living in this vast area travel for their corn in the same way that the sons of Jacob did into Egypt, and they drove their flocks of sheep in the same way for pasture.

Our way to Nan-gutty lay over mountains, forest, creeks and valleys and wooded areas covered in vines and my luggage was very troublesome as I had large cumbersome packs containing birds and curiosities to carry.   We stopped here the night and set off for Bundi which meant crossing the Morris Mountain, twelve miles across and the Manara plain, most of the way following a marked treeline. A very rugged trail it was, but we managed to reach Bundi in the evening.

Here I met up once again with he whom I termed "Buttermilk" John Ashmore doing other people's work and neglecting his own.   Here at Liscome's place were my cart and boxes.   It will be recalled that Liscome had gone to Sidney with all my money from the box and my tarpaulin.   It seems that before leaving he had offered Ashmore ten shillings a week to stop at his place until he returned from Sidney although, he and Ashmore well knew that he was contracted to me and that I was expecting him at Captain Stephenson's.

I stayed here a fortnight but Liscome did not return so I went to Manara, Thomas Warburton's place for a few days shooting and walking until I learned that Liscombe had passed by Manara on his way to Bundi during the night.   I set off after him on my horse and when I came into the yard there I saw my tarpaulin cut to pieces lying on the ground.   I travelled on to Liscome's with some blacks catching squirrels on the way. When I arrived I was told that Liscome's "tally wife" was drunk in bed having freely imbibed of the liquors that he had brought back from Sidney. I found Ashmore my Man Friday in the milking yard putting some calves up. Whilst we were talking my dog and Liscome's began fighting over a piece of kangaroo meat and both ran into the passage, meanwhile Liscome was holding down his tally-wife who was now both drunk and mad.

Then Liscome ran out gun in hand, threatening to shoot my dog and me too. We scuffled and at that moment his tally came staggering out and somehow managed to get bitten by my dog. Liscombe gathered her up in his arms and took her in the house. Ashmore and I parted the fighting dogs and everything quietened down.

As I was very hungry having had nothing to eat all day we went into the kitchen to be challenged by Liscome who demanded to know what we were doing there and claimed that we had come to rob him. Threatening to shoot us both he ordered us from the house, so we went out and spent the night with the blacks who were camped outside, I lay with my loaded guns beside me, determined that if he made a move against us I would shoot him dead.

In the morning without breakfast or supper the night before, we loaded up the cart and set off for Thomas Warburton's at Manara and stayed a few days before reaching my old master's place at Bonbarlow. I stayed with him a fortnight during which time Ashmore and I parted company.

Soon afterwards I met with two drays drawn by bullocks coming down the country. The draymen told me that they were on their way to Goulborne and they agreed to carry me and my luggage there. Three days later, having slept each night under the drays, we came to Rock-flatt where there is a well which contains water of a wonderful nature, excellent to accompany spirits and, if used for kneading flour it raises the paste quicker than any yeast, moreover the cattle are particularly fond of it.

Further on we arrived at Jew's Flat where there is a public house at which we had to pay four shillings a quart for racked ale which nevertheless was very welcome.

I never suffered so much in my life as I did on this journey from the dry, hot, dusty winds. Although we had taken about four gallons of water on the journey with us it was finished before we had gone half way. We would have perished had we not met two gentlemen in a gig who gave us each a gill-glass full of brandy. We stayed at Jew's Flat for a day and for the next two we travelled over mountain, forest and bush until on the third day we came to the Tindry Mountains which are very beautiful, stretch for many miles and are inaccessible to man. In this area there are both wild cattle and wild pigs.

Next day our way passed over rough barren rocks where we often had to hold the drays back with ropes until we arrived at the top of Limestone Plains. Reaching the end of these the following evening, we found a fine station that belonged to Captain Merchant Camel and we stayed the night. Over the following two days we had to take to the bush again until we reached the foot of a two and a half miles high range which we tackled the next morning, needless to say it was of course the same distance down the other side. We had to fell some timber at the top to act as a drag when we descended. That night we came upon Lake George.

I was told that thirty years before, this lake which is twenty one miles long and ten across would sail quite a large ship but was now completely dry, as level as a bowling green and well stocked with all kinds of cattle."

Lake George is known as the Disappearing Lake, a place of myth and legend about 40km north of Canberra. It is not fed by any rivers or streams being completely dependent upon rainfall. A string of wet years results in a full lake, which dissipates quickly again as the hot sun and a series of dry years empty it.

Joseph goes on "The next day we arrived at Mandervin Plains where there are many kinds of beautiful birds to be seen.   Then into the bush again arriving the same night at Goulborn. Here I left the men and drays as they were headed to Bong-bong and I travelled another ten miles to Mr Camel's, arriving three days before Christmas Day.   I stopped with him for five months assisting at the harvest and had some more shooting until I had filled up my cases and shot no more."

Chapter 9

At the end of May, Joseph travelled the hundred and twenty miles down to Sidney and met Mr Sparks again, who with four other men owned a store in Blithe Street and suggested that Lingard joined them. He agreed and moved in to live there for three months until unfortunately the business failed and they went their separate ways.

Joseph tells us that at this time things were very bad in Sidney, work was impossible to find, his money was exhausted and he could not get a passage to England. So desperate was he for money that he was forced to sell his favourite gun which was quite a wrench, and twice, he records his cases and luggage were impounded against unpaid rent by landlords where he was lodged at different times. He declared that the whole town was in a state of distress with eighteen hundred unemployed, and the best of mechanics who were formerly earning twelve shillings a day were, if they could get the work, getting ten shillings a week. Most of the banks had failed and property generally was of little value. He instances that a "good bullock" could be bought for a pound, and sheep for two shillings each. Twenty guinea watches changed hands for a bare fifteen shillings.

He goes on "As four ship loads of people left Sydney to South America to seek work, three shiploads of immigrants came into Sydney from England, Ireland and Scotland. One vessel brought chiefly women. I saw them land on Camel's Wharf and a few days later I saw some of them crying at the ends of streets neither a penny in the pockets, a mouthful of food to eat nor friends to look after them.

I have seen them getting up in the morning from under the wild trees not having had a farthing to pay for lodgings, others begging the bullock drivers going up country to take them along to do what they will for the sake of a morsel of meat.

Many hundreds of people there who wished themselves back in England, had not a penny to bless themselves with, that included myself – I was almost at my wits end" he declared.

Joseph goes on to say that on learning of his distress, Captain Camel his former master, sent for him to come and stay for as long as he liked with him at Golbourn, but declined to accept his kindness as he was fully set on returning to England.

"At length providence raised me up a friend" he said.   It appears that a man that had been transported like Joseph, now worked as a ships glazier and painter and was a great friend of Captain Wattle of the barque *Aden* that had recently put in from London.   He took Joseph aboard and through his influence, and a character reference from Captain Camel he was taken on as assistant cook and steward of the ship.   All his cases and boxes were put on board under the instruction of the Captain where they would stay until Lingard could pay him ten pounds for their freight to London.

Joseph goes on in his own words "When the *Aden* was freshly loaded with three hundred tons of sugar, tea and other cargo for Port Philip I went aboard on the 25th of January 1844 and on the 26th we sailed out of Sydney harbour.   From Sydney to Port Philip is about seven hundred miles by sea through the straits betwixt Van Dieman's Land and the mainland.   On the 30th we arrived at Port Philip, the land around which is low, level and better for cultivation than in the vicinity of Sydney.

Five mile from Williamstown up the River Yarrow is the town of Melbourne, which is the principal place within three or four hundred miles. At the port were several vessels loading with wool and bark both of which are transported from the interior by carriage. The bark is beaten flat with hammers and stowed fore and aft in the hold creating dust almost to the point of suffocation for everyone nearby. When the wool comes on board in large bales they are pressed down until three take up the space originally taken by one.

Whilst we were at the quay a ship the *Sir William Wallace* from Liverpool arrived with immigrants and out of the three hundred and forty men, women and children who left England after a bad voyage only three hundred and one arrived at Port Philip. It was not until the 30th of March with all our cargo loaded, that we eventually left for Old England. The weather as we sailed by Van Diemans Land was very fine and we saw the lady passengers skipping about on deck anticipating a fine voyage. Soon after we came in sight of New Zealand and were becalmed for a week. Then we came into the South Seas and with winter approaching fast, we had heavy storms almost daily. Those same ladies had given up their mirth and dancing and were in a sad condition below through sea sickness."

Joseph goes on to describe the rest of his return voyage. "We passed the South Sea Islands and rounded Cape Horn on about the 20th of June the shortest day there. Here we were amongst icebergs floating on the water, some several hundred feet high and were a month sailing through them. We passed the River Plate and also the island noted as being the residence of Robinson Crusoe.

Here we encountered what is called a black squall which came close to disabling the ship, snapping the main yard and fore yard each into three pieces and twisting the boom off by the forecastle.
 At this point Colonel White, a passenger, who had been ill since we left Port Philip, having lost the use of his limbs and speech, died. All hands were mustered aft, with the bell tolling, his body was brought out sewed up in a hammock with sandbags at the feet and laid on grating with the Union Jack spread over it. The service was read and the body committed to the deep, as the grating was raised the body slid off and fell with a great splash into the sea, when to our horror the corpse rose to the surface some distance astern

Shortly we came to Rio de Janeiro, which I think is the finest harbour in the world.   On the starboard side is the Battery and on the larboard Sugar Loaf Hill, which is of such a great height it would surely be an impossibility to get to the top. We passed other batteries for a few miles and then dropped anchor opposite the town.   The governor's castle stands on a hill above the town and the surrounding country is very hilly some looked almost perpendicular. Rio is a place of great trade with merchants coming from all over the world, dealing mainly in cotton, sugar, coffee, oranges, lemons and spices of all sorts.

 I was informed that in that country there were two and a half million slaves who are bought and sold in the markets like cattle. We stopped here for four days taking on water and provisions and sailing away on the Sunday morning.   We pursued our way until we crossed the line and hereabouts there were sea birds of all descriptions, flying fish, dolphins and sperm whales all close to the ship.

We next came to an island belonging to Portugal. In appearance it was just a barren rock some nine miles long three and a half broad with about ten thousand inhabitants. The interior abounds with corn, vines & etc. We took on here a live bullock and plenty of fowls at three pence each, plus four tons of water.
We could have exchanged any quantity of wine bottles at the rate of two bottles for one hat full of eggs. However, there was no anchorage to be found here and we tacked about the front of the rock all day whilst taking in our cargo. The men and women living here are the very colour of a tallow candle, but the females are not to be compared with the rosy-faced girls of Old England."

He notes that they were now about eighteen hundred miles from England and as they set sail his excitement and anticipation became more intense just to catch a glimpse of his native land. About two o'clock in the morning on the 24th of August, the sailors on watch called to him on deck to view a lighthouse on the shore of England. He writes "I obeyed the summons with great pleasure and was instantly on deck in nothing but my shirt. That same day we reached the Channel and steamers ready to tow us up to Gravesend. There we engaged another pilot and steamer to tow us up to London Dock where we arrived safe on 30th of August 1844 having not set foot on land since Sydney on the 25th of January. I went ashore with the cook, the butcher and a few friends to get a pint of porter in Ratcliffe Highway.

I stayed in London a fortnight before I could get my boxes and other luggage clear of the Customs House and then took the railway to Manchester, meeting two friends there who accompanied me to Chapel-en-le-Frith. Then I carried on nearby to my native village after an absence of nearly ten years."

Joseph ends his memoir with a poem:

Sweet home whilst on a foreign shore,
I often thought of thee,
And sighed lest I should never more
My wife and family see.

The flowers that bloomed around me there,
To me they seemed not sweet,
I longed to sit in my old chair,
My children's smile to meet.

But though Old England - lovely spot -
I dearly wish'd to see,
Alas! Deserted was my cot,
I found no home for me!

T.B.

Epilogue

In 1844 when he arrived back in England after his enforced exile, Joseph Lingard was aged 54 and obviously stayed in or around his native village for at least two years as his little book was published at Chapel-en-le-Frith in April 1846. However he appears in the 1851 Census, lodging in Stockport, Cheshire with a widow who had been born in Chapel-en-le-Frith and may have been a relative or an old friend. He is shown as a widower in the same census although in his book there is no mention made of a wife or family.

The plaintive little poem he signs off with, might be taken to indicate that his wife had died before or during the time he was away and the house he had lived in at the hamlet of Bridgeholme Green, near Chapel Milton was empty but despite searching the records I have as yet been unable to trace anything.

There have been Lingards living in and around the Chapel-en-le-Frith area of Derbyshire since at least the 16th century and probably earlier, since the name probably originates in the Danish and Norwegian name Lingaard, which translates from the ancient Nordic language as "farm or open space with lime trees." In Joseph's time there were still several Lingard families living and mostly farming in the district although many who did not own their own farms or were working as farm labourers had moved to work in Lancashire in the booming cotton trade mostly in and around Ashton-under-Lyne, or over the Cheshire border in Dukinfield.

Whilst Joseph, as far as I can trace, is not in my direct line of Lingards, without doubt we have a common ancestor perhaps seven or eight generations back, and having shared his memories of what happened nearly two hundred years ago I salute his courage, fortitude and determination and am proud to carry his name.

Denys **Joseph** Charles **Lingard.**

July 2015

Acknowledgements

As always with my writings I must first thank my wife Mary for her unstinting encouragement, suggestions and corrections, similarly my daughter Sally and lastly my grandson Sam whose patience and help in computer matters have been invaluable.

This book was never intended to be a scholarly study of 19th century justice, convict hulks, transportation or indeed life in Australia in the years of early settlement, and I owe much in my background studies to two books in particular. Both are eminently readable, well researched and are premier works in their fields - *The Intolerable Hulks* a history of British Shipboard Confinement 1776-1857 by Charles Campbell (Third Edition 2001 Fenestra Books, Tucson, Arizona, USA) and *The Fatal Shore* the formidable History of the Transportation of Convicts to Australia, 1787-1868 by Robert Hughes (1987 Collins Harvill, London).

Other sources to whom I acknowledge my indebtedness for help and guidance are – The British Library, The National Maritime Museum, The National Library of Australia, and the State Library of New South Wales.

Denys Lingard

About the Author

Denys Lingard was brought up in Cheshire and was educated at Stockport Grammar School.   Resisting family pressure to read Law he approached the editor of a local newspaper whom he persuaded to appoint him as a junior reporter until he was called up into the RAF from which he was 'demobbed' in 1949.

New opportunities beckoned in London and he absorbed courses on sales, sales management and the then new concept of marketing. He became national sales manager of a quoted company in his thirties and a director before forty.   When this company was taken over by a larger multi-national organisation his responsibilities grew and his work then entailed a good deal of international travel.

Married with three children and five grandchildren, Denys is retired and lives with his wife Mary in the Cotswolds.

His first novel – a mystery thriller *"The Beautiful Obsession"* was first published in 2012.

CPSIA information can be obtained
at www.ICGtesting.com
Printed in the USA
LVHW081604041019
633218LV00029B/688/P